THE STINSON BEACH SALT MARSH

The ground fog and high tide had effectively isolated the grass.

Only the ends of the stems and the blooming tips
broke the surface of the high water around the edges
while the center of the stand was completely inundated,
forming a hollow.

Even with the ducks beyond
and the sprouts of Bulrush in the foreground
that stand of Cord Grass for a long moment
was totally set apart in space and time
and I along with it.

"It's a sewer," said the child.
"It's a salt marsh," said I.

This work is for Betty
whose encouragement and help made it possible.

Library of Congress
Catalog Card Number 77-70990
ISBN 0-918540-01-1

Published at Stinson Beach
Stinson Beach Press
P.O. Box 475
Stinson Beach, California 94970

THE STINSON BEACH SALT MARSH

The Form of its Growth

Photographs
and Text
by

Bernard Poinssot

Stinson Beach Press
1977

There's a lot going on in the Stinson Beach Salt Marsh.

If you watch it long enough you know that it's never the same.

But it takes a special kind of looking
and maybe imagination as well to find out why.

Mostly it takes waiting and watching days in and out over a long enough cycle
and from all around the edges and the hills above in fog, full sun and rain,
as the sun comes up and at night and especially at all tides and seasons.

It also takes a certain kind of obsession.

In any case it did for me.

The marsh along Highway 1 north of Stinson Beach
with the Bolinas Lagoon beyond.

It all started quite early one morning in the fall
at a high tide and in a very heavy fog.

I saw a pure stand of Cord Grass
which I had passed and noticed a lot before
but never as it appeared then
in that flat light.

The place suddenly became
and still remains a center of mystery
and a focus in the marsh to which I always go.

The ground fog and high tide had effectively isolated the grass.

Only the ends of the stems and the blooming tips
broke the surface of the high water around the edges
while the center of the stand was completely inundated,
forming a hollow.

Even with the ducks beyond
and the sprouts of Bulrush in the foreground
that stand of Cord Grass for a long moment
was totally set apart in space and time
and I along with it.

After that I came back a lot.

At first only to that same Cord Grass
to understand and to learn more about the magic of the place,
but very soon I discovered that it was not only the Cord Grass
but the whole salt marsh which was enchanted and from then on
I was very involved in trying to figure out what was going on.

And why.

Bernard Poinssot
Stinson Beach, California
June 30, 1976

CORD GRASS: Winter 1974

THE STINSON BEACH SALT MARSH

The Form of its Growth

CORD GRASS

I found out that Cord Grass,
Spartina foliosa and a true grass,
is a pretty important part of the marsh even though,
compared to one time when there must have been more,
it now occurred infrequently as a large pure stand.

From my original place it trailed away
partially submerged like rice in a paddy
eventually becoming, with stands of Bulrush,
a narrow fringe along one edge of the marsh
coinciding more or less with the average high tide line.

It also flourishes along the other side of the marsh
bordering the highway mixed in along the shore
and in small stands of interlocking ovals and circles
twenty to thirty yards from the bank.

Finally there is one major stand
near the middle of the marsh lying in a circle
just above and next to a bend in the main tidal channel
winding between the creek and the lagoon.

At low tides this relationship can be seen quite clearly.

At a high tide from the hill above late in the day
for a short while that particular stand looks dark
and totally isolated above the shimmering flooded marsh
until the waters recede, the mud banks come back into view
and the birds and water fowl arrive in great numbers
to feed on the flats and to bathe in the shallow sloughs.

At close range the effect of the grass is something else.

The individual culms, almost two feet high
and cutting up through the water in groups of four and six,
are connected by a dense network of rhizomes in the mud
which form a solid anchorage system.

They stand erect although a light wind makes them shiver.

In late August generally,
scaly flowers appear as barbed spikes
at the ends of gradually narrowing stems.

In October and November the stems are still green
but the spikes become golden yellow tips
and create still another dimension
since flowering is much more dense around the edge
than in the middle of a stand.

Even though the fuzzy tips turn brown in late November
the overall effect of a stand of grass is still green
except that through the maze of undergrowth
and in areas where the grass is thinning out,
where long legged black and white water birds
thread and pick their way,
the blackened stumps of Cord Grass in the dark mud flat
are like the charred remains of an old fire.

Among all the marsh plants
Cord Grass is visibly unique in its ability to withstand prolonged submergence,
and apparently no other plant produces as much oxygen during its lifetime
nor as much nutrient material when it decomposes into tiny particles
as it does gradually and constantly throughout the year.

BULRUSH

Another early discovery was Bulrush.

Scirpus robustus, of the Sedge Family to be exact,
grows along the border with Cord Grass
in thick dense stands in two or three places
but more often in thin trailing meanders
at the margin with Pickleweed, Salt Grass and Jaumea.

I first really looked at it closely in late June
when the yellow green triangular stems stood erect
and healthy about three feet high
ending in a tuft of three unequal sharp leaves
which cradled a tight cluster of small flowers—
reddish brown, fuzzy oval spikelets.

Below, in the salty mud, the Bulrush was spreading
by sending out horizontal stems
thick with a reserve food supply
to produce more stems above
and a broader root system below.

CORD GRASS: Winter 1975
Winter 1975
Winter 1975
Winter 1974
Winter 1975
Winter 1975
Fall 1975
Fall 1975
Winter 1975
Fall 1975

In the middle of September the stems start to turn.

*By November they are weathered to a gray brown,
getting dry, caked in mud and
broken by wind, dying to the ground.*

*The previous year's growth lingers on
but by April, beaten by the rain, is almost completely gone
or hard to distinguish in the complex of decayed stems
through which the first new shoots start to appear.*

*The new growth does not always follow old patterns
causing subtle and dramatic changes in form and contour
all along the margin.*

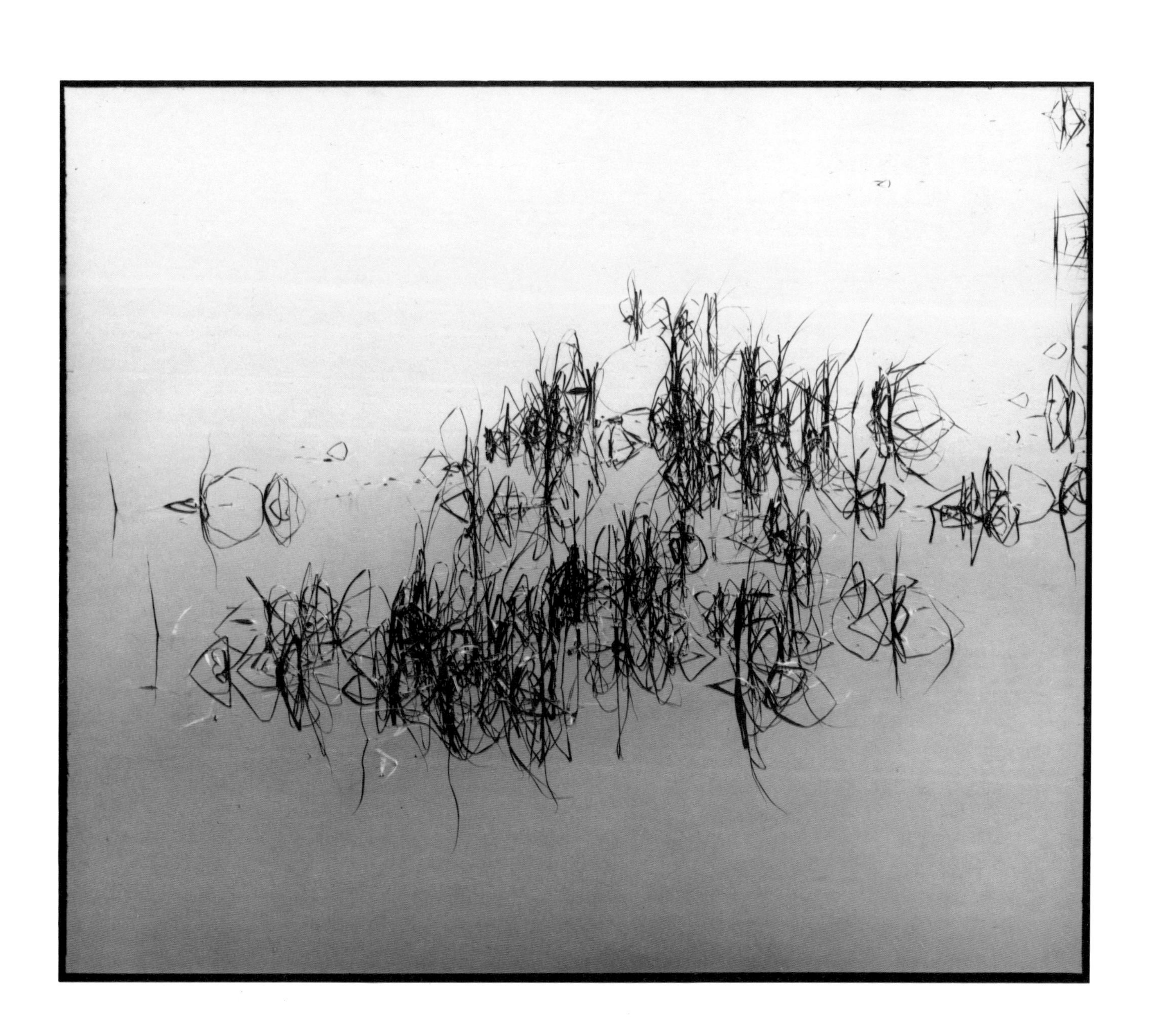

BULRUSH: Summer 1975
Summer 1975
Fall 1975
Fall 1975
Summer 1976
Winter 1975

TULE

*Of the Sedge Family, like the Bulrush, is the Tule, Scirpus rubiginosus,
a giant Bulrush which may have originated as a coastal form hybrid of two similar species.*

*It grows as high as fifteen feet, straight up or in a slight arc set by the wind,
as a single perfectly round stem up to an inch across as it comes through the mud
tapering slowly to a spear-like point at the tip
which supports a bristling cluster of red brown spikelets.*

*It is found growing densely and continuously for 250 yards only
along a bank near the highway in the north easterly part of the marsh
and in a long neat stand between tidal channels paralleling the bank.*

*Where it begins to thin out the closest other marsh plants
are Cord Grass on the deep water side and Pickleweed toward the shore.*

*The stems start out in April
a rich healthy looking yellow green
with occasional brown flecks.*

*As the season goes on the stems turn yellow–
the spots increase and get larger–then brown
and finally the gray white of weathered wood.*

*They break and fall creating a jungle
like some unique basketwork structure.*

*Tule stems were used by the Native American tribes
to weave mats and to build their balsas or small boats.*

THE MARGIN

The margin is the band around the marsh
which just about agrees with the average high tide line
where Cord Grass and Bulrush begin to thin out
and other plants such as Pickleweed, Salt Grass and Jaumea,
more suited to that particular salt water environment
take hold and start their strong growth
and provide the dominant form, texture and color.

The zone varies from a yard wide
to several hundred in one particular place
at elevations of 5.1 to the highest tide of about 6.6 feet.

It is a fascinating area
because it is constantly changing,
there is a lot of variety
and since it is highly visible at close range
it is possible to observe clearly
the effects of the flooding and ebbing,
the alternate wetting and drying
and the recycling of nutrients and waste
by the cyclic rhythm of the tides and the sun.

TULE: Winter 1975
Summer 1976
Fall 1975
Winter 1975

PICKLEWEED, SALT GRASS AND JAUMEA

Pickleweed (Salicornia virginica),
Salt Grass (Distichlis spicata)
and Jaumea (Jaumea carnosa)
are usually, or at least more than half the time,
found in that order as the land rises in the margin.

At other times Salt Grass is first at the edge of the water
and occasionally there is a mat of Jaumea
spread much farther out than usual,
the first to be flooded by high water
and the last to be exposed by the falling tide.

There is probably more Pickleweed
than any other marsh plant.

Its tangle of succulent and leafless horn-like jointed stems
forms a basic ground texture
against which other plants are seen or not
depending often on the season and the location.

It has a salty taste and can be eaten as a fresh vegetable.

DRIED PICKLEWEED: Winter 1975
PICKLEWEED AND CORD GRASS: Summer 1975
PICKLEWEED AND BULRUSH: Summer 1975
BULRUSH AND CORD GRASS: Fall 1975
PICKLEWEED: Winter 1974
PICKLEWEED, BULRUSH AND CORD GRASS: Summer 1975

Salt Grass has an entirely different structure
and surface quality.

Of the Grass Family, like Cord Grass,
its stiff wiry culms with pointed leaves
grow mostly erect or in an ascending line
in dense irregular patches.

Its flowers, pale green
drying to straw color in late summer
are in the form of small spikelets
growing generally at the tops of culms
in clusters up to an inch and a half long.

From June to November
the yellow spots of color in the margin
are the small disc flowers of Jaumea
which instantly identify it.

A closer look
will reveal its smooth fleshy leaves
and the matted and horizontal nature of its growth.

Pickleweed must be considered a dramatic marsh plant
but it is capable of modest effects as well.

In July, briefly and unevenly,
the appearance of yellow dust is seen under a glass
to be tiny flowers in groups of three in the joint of the stems.

The dazzling spectacle is reserved for the end of November in the low sun of winter
when the broad colonies of the succulent stems become a red the color of crabapple
made transparent by the light coming from behind and tinged with yellow and orange.

COTULA

Growing occasionally in the margin near Pickleweed,
Cotula coronopifolia,
also called Brass Buttons,
is somewhat different
because it is relatively uncommon
and is considered to be almost aquatic.

There are not many plants in the margin
growing very low and almost always in wet ground.

The only other place it seems to flourish
is in a long line running along a mud bank
on one side of a tidal channel,
an area which is subject to prolonged submergence
even at moderately high tides,
and which does not seem to support any other plant life.

Of the Sunflower Family, like Jaumea,
the small, flat and round flower heads,
blooming nearly all summer
are bright clear yellow discs
at the end of many branched smooth stems
somewhat succulent, as is common in the marsh,
but not as fleshy as Jaumea or Pickleweed.

The yellow green leaves,
which start out at their base
as a sheath around the stem,
take several shapes in the same plant
from gently toothed or lobed
to the flat perfect form of a lance.

PICKLEWEED IN BLOOM: Summer 1976
DRIED PICKLEWEED: Winter 1975
JAUMEA IN BLOOM: Summer 1976
SALT GRASS AT HIGH TIDE: Winter 1975

ARROW GRASS

The south easterly part of the marsh,
where the margin is at its widest and highest level,
is where all along the high edge
Arrow Grass–Triglochin maritima–
seems to grow best in scattered clumps up to a foot wide
mixed in an area where the main growth is Salt Grass
with occasional spots of Sea Lavender.

The leaves of the plant,
which are long and very slender,
are in the shape of a crescent
in cross section, not flat
and spring forth at a low angle
in a dense tuft
from a short stout root stock.

From April until August
small greenish flowers bloom densely
along a single axis for three inches
at the end of a thin flowering stalk
up to two feet long and limber enough
to wave in even a light breeze.

Another species of the Arrow Grass Family
is Triglochin concinna
which also grows in the upper margin
although not as plentifully
and differs in the cross section
and shape of its leaves,
which are almost round and more slender,
and in the wider spacing of its blossoms
along the flowering stalk.

ABOVE HIGH TIDE

As the ground rises the short distance between the average high tide
and the highest tide, from about 5.1 to 6.6 feet,
the growing condition changes dramatically
from one of frequent and prolonged salt water submersion,
ideal for plants in the margin–Pickleweed, Salt Grass and Jaumea–
to dry land or only occasional wetting at high tides
or at times of high water combined with wind
and the result is a relatively narrow zone
which supports an assortment of different marsh plants
varying greatly in their flowering and growth habits
and thriving primarily in the limited elevation range
from 6.6 to 8.0 but also up to elevation 10.0 feet.

Exceptions are frequent
and quantity and density are not uniform
but generally Sea Lavender and Frankenia
are found at the lowest elevations above high tide
followed by Saltbush and Grindelia.

COTULA IN BLOOM: Summer 1976
ARROW GRASS IN BLOOM: Summer 1976

Sea Lavender or Marsh Rosemary–
Limonium californicum–in flower,
is always recognized as a single isolated plant
with an unusual combination of leaf blades and flowering stems
seen against the smaller scale background texture
of Pickleweed, Salt Grass and Jaumea,
at the lower elevations of its growth
and Frankenia at somewhat higher ground.

There is no other plant quite like it in the marsh.

The broad, flat, oval shaped leaves,
up to eight inches long,
spread in a clump from a thick woody root
and, except for their large size,
are relatively inconspicuous until June
when a dry looking and angular branching stem
grows from the center of the clump of fleshy leaves.

The leafless flower-bearing stem rises quickly to two feet
and in July small pale violet-purple flowers
mature in clusters at the tips of the many branchlets
creating sprays of color
in strong contrast to the basically green background.

The flowering is both subtle and extravagant
all through the summer.

By early October the small lavender flower spikes
which always had a transparent and fragile quality
turn a creamy white and in another month are brown
along with the branching stems.

In contrast to Sea Lavender,
growing in small spreads as well as large masses
in the same general area but usually higher,
is Frankenia grandifolia,
an abundant low growing and bushy perennial
which has small gray-green leaves
clustered on short stems
in an intertwining growth habit
six to twelve inches high.

Its small pink flowers
bloom in great numbers
from early June to late October
and provide fine texture and color
in a clearly identified horizontal zone.

Saltbush (Atriplex patula var hastata),
which grows inconspicuously and in spots
on land well above high tide in the westerly part of the marsh,
is a small drab annual which belongs to the Goosefoot Family,
the same as Pickleweed, but it is not nearly as interesting.

It branches out up to two feet in all directions flat on the ground
and flowers late in the year, from the end of August through November,
a small reddish spike from between distinctive leaves the shape of an arrowhead.

At the end of the blooming season
the edges of the leaves and the branches turn a rose color
adding to the dazzling spectacle of Pickleweed turning at the same time.

Gum plant (Grindelia humilis),
belonging to the Aster Tribe
of the Sunflower Family,
grows plentifully
all around the marsh
and, it seems,
equally well
in formation
in long lines
marked
by the upper reaches
of the tides
or in dense clumps
at higher elevations.

It is a bushy shrub
reaching four to five feet
in many places.

The light green leaves,
up to three inches long
and with saw toothed edges,
are oblong shaped
but pointed at the tip
and tapered to the base.

Before opening
the flower heads
are white with a gummy secretion
but in bloom, from June
through the end of October,
the bright yellow ray flowers,
two inches across,
are the most striking points of color
around the entire salt marsh and beyond.

When this work was in its final dummy stages
it was shown to a number of people
and those whose reactions I benefited from most
were the writers Harold and Ann Gilliam and Jerry Mander,
the paleontologist Tom Williams,
the publisher Frederick Mitchel,
and Sterling Bunnell whose fields are psychiatry
and plant and animal ecology
and who lifted my spirits more in an hour
than anyone else has in a lifetime.

B. P.

SEA LAVENDER IN BLOOM: Summer 1976
GRINDELIA BLOSSOMS AND BUDS: Summer 1976

This book
was printed
in an edition
of one thousand
five hundred copies
by Cal Central Press on
Scott-Warren 100-lb. Cameo
Gloss paper and 12-point Lustercoat
cover. The photographs were reproduced
by double impression offset lithography using
300-line screen duotone plates from custom prints by
Irwin Welcher. The text was composed in Palatino photo type
by Robert Sibley. The Smythe-sewn binding was produced by Cardoza-
James Book Binding Co. The design and layout were done by Bernard Poinssot.

BIRDS FEEDING IN THE MARSH: Winter 1975